READING POWER

Man-Made Disasters

Toxic Waste

Chemical Spills in Our World

August Greeley

The Rosen Publishing Group's
PowerKids Press™
New York

Published in 2003 by The Rosen Publishing Group, Inc.
29 East 21st Street, New York, NY 10010

First Edition

Book Design: Christopher Logan

Photo Credits: Cover © Peter Turnley/Corbis; pp. 4–5 © Richard Hamilton Smith/Corbis; p. 5 (inset) © James P. Blair/National Geographic Image Collection; pp. 6, 9 (top), 11, 12 (maps and illustrations) Christopher Logan; pp. 6, 12, 17 (globe logo) © PhotoDisc; p. 7 © AP/Wide World Photos/ Effingham Daily News; pp. 8–9, 14, 17 © AP/Wide World Photos; p. 10 © Chris Rainier/Corbis; pp. 10 (inset), 16 © Reuters NewMedia Inc./ Corbis; pp. 13, 15 (top and bottom) © AFP/Corbis; pp. 18–19 © Midwest Treaty Network, www.treatyland.com; p. 20 © Joseph Sohm, ChromoSohm Inc./Corbis; p. 21 © Todd Gipstein/Corbis

Library of Congress Cataloging-in-Publication Data

Greeley, August.
Toxic waste : chemical spills in our world / August Greeley.
     p. cm. — (Man-made disasters)
Summary: Explores how dangerous chemicals can be released into the environment, the harm they can do, and what can be done in the future to prevent such disasters as the release of cyanide into Europe's Tisza River in 2000.
Includes bibliographical references and index.
ISBN 0-8239-6483-3 (library binding)
1. Chemical spills—Environmental aspects—Juvenile literature. [1. Chemical spills—Environmental aspects. 2. Hazardous wastes—Environmental aspects.] I. Title. II. Series.
TD196.C45 G74 2003
363.738'4—dc21
                                                      2002000508

Contents

Chemicals in Our World

People use chemicals to make life better and easier. Some chemicals help plants grow faster. Others keep bugs off of crops. Some of these chemicals are toxic, or poisonous. When used carefully, toxic chemicals can be very helpful.

This plane sprays chemicals on farm crops. If not used carefully, chemicals can pollute the air, land, and water.

However, when toxic chemicals get into the water, land, or air, they can harm the earth. The most toxic chemical spills can kill plants, animals, and even humans!

The waste from a chemical plant in Russia polluted the air and water for miles.

Each year, there are hundreds of harmful chemical spills around the world. Spills can happen for many reasons. Sometimes, ships, trucks, and trains carrying the chemicals crash. Other times, parts of factories may break down, allowing the chemicals inside to spill outside.

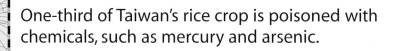

Check It Out

One-third of Taiwan's rice crop is poisoned with chemicals, such as mercury and arsenic.

Russia

China

India

TAIWAN

Many chemical spills happen when the chemicals are being moved from place to place. Chemicals were spilled in this train accident.

A chemical spill does not take long to harm living things. A spill can quickly poison small plants and animals. These plants and animals may be eaten by other animals. The toxic chemicals in the plants or animals can harm the creatures that eat them. A single spill can harm or kill thousands of living things.

There was a chemical spill in a Florida lake about 20 years ago. Poisons from those chemicals still harm animals in the lake today!

4 Birds and people eat big fish.

3 Big fish eat little fish.

2 Little fish eat algae.

1 Algae

The toxic chemicals that are taken in by algae are passed on to others in the food chain.

WARNING Chemicals in Silver Lake may be harmful to humans

The Community Right to Know Law

In 1984, there was a huge chemical spill in India. It killed 3,000 people and 100,000 others were hurt. Two years later, the Community Right to Know Law was passed in the United States.

The chemical spill in a plant in India angered many people.

The law says that companies must tell the public what chemicals are being used in their factories. Every year, companies must also report the amount of toxic chemicals that they are letting out into the air, land, and water.

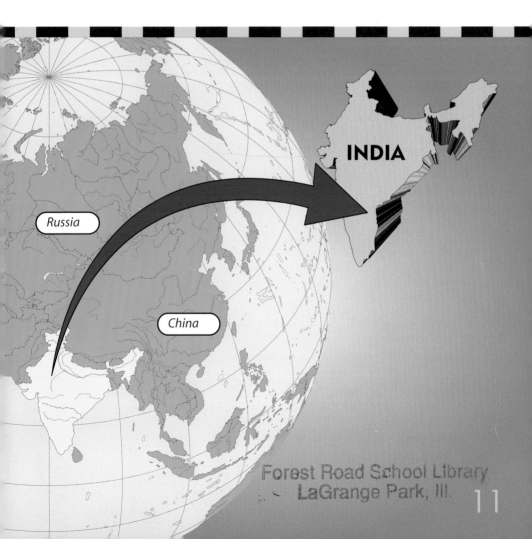

INDIA

Russia

China

Chemical Spill in Romania

Since the Community Right to Know Law was passed, people have become more aware of the dangers of chemical spills. These toxic spills are still a problem, though.

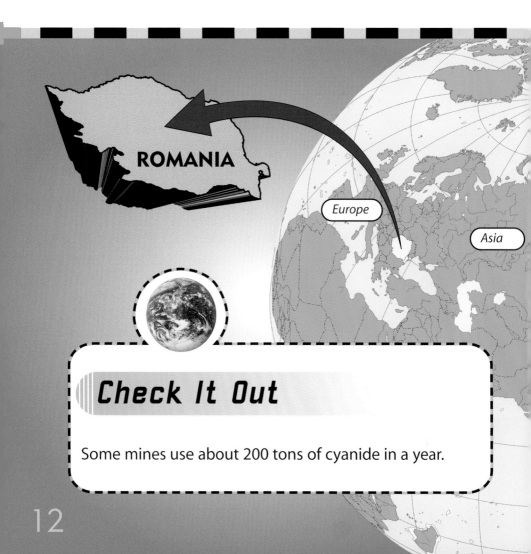

ROMANIA

Europe

Asia

Check It Out

Some mines use about 200 tons of cyanide in a year.

One very big spill happened recently.
A gold mining company in Romania used
a chemical called cyanide to separate
gold from ordinary rock. The mining
company had built a dam to keep the
cyanide out of the surrounding land
and water.

*The owners of the mine in Romania did not
tell people about the spill for 24 hours.*

On January 30, 2000, heavy snowfall caused the dam to overflow. Cyanide spilled into a nearby stream. The polluted water flowed quickly from the stream in Romania into Hungary's second-largest river, the Tisza.

Before the spill, the Tisza River was one of Europe's cleanest rivers!

Many living things in the river, including bacteria, birds, and otters, were killed. More than 1,200 tons of fish died from this man-made disaster.

Many people worked to clean up the cyanide spill in the Tisza River.

People quickly pulled the dead fish from the Tisza River. They did not want other animals to eat the poisoned fish. About 15,000 fishermen lost their jobs because the river was so polluted.

About 26.2 million gallons of cyanide and polluted water were spilled into the Tisza River. It may take more than 10 years before fish can live in the river again.

Check It Out

Poison from the chemical spill in the Tisza River left 2.5 million Hungarians without water to drink!

Some important lessons have been learned from the spill in Romania. People who lived in Wisconsin saw on TV what happened in Hungary.

They got their state government to pass a law saying that cyanide cannot be used in Wisconsin mines.

People in Wisconsin marched against using cyanide in mines in their state. They did not want a spill to happen near their homes.

⊸Working Together

Businesses that make and use toxic chemicals are working with their communities and governments. They are trying to make chemicals that are less harmful to the earth. They are also doing everything they can to prevent toxic spills and keep the earth safe.

Toxic chemicals must be gotten rid of safely so that they do not pollute Earth.

On Earth Day, many people gather in Washington, D.C. They want to let their leaders know how they feel about harmful chemicals.

Glossary

algae (**al**-jee) small plants without stems or leaves that are found in water

arsenic (**ahr**-suh-nihk) a very poisonous chemical that is sometimes used to kill bugs and weeds

bacteria (bak-**tihr**-ee-uh) very tiny living things; some can be useful, while others cause sickness

chemical (**kehm**-uh-kuhl) a substance that is made of elements such as carbon and oxygen

chemical spill (**kehm**-uh-kuhl **spihl**) when toxic matter harms land, water, or living things

cyanide (**sy**-uh-nyd) a poisonous metal salt used in mining and in the making of plastics and bug sprays

disaster (duh-**zas**-tuhr) a sudden event that causes great loss or harm

mercury (**mer**-kyuhr-ee) a silver-white element that is poisonous

poisonous (**poi**-zn-uhs) deadly

pollute (puh-**loot**) to make the environment dirty with man-made waste

toxic (**tahk**-sihk) harmful or poisonous

Resources

Books

Chattanooga Sludge: Cleaning Toxic Sludge from Chattanooga Creek
by Molly Bang
Harcourt (1996)

Love Canal: Toxic Waste Tragedy
by Victoria Sherrow
Enslow Publishers, Inc. (2001)

Web Sites

Due to the changing nature of Internet links, PowerKids Press has developed an on-line list of Web sites related to the subjects of this book. This site is updated regularly. Please use this link to access the list:

http://www.powerkidslinks.com/mmd/twch/

Index

B
bacteria, 15

C
chemical, 4–9, 11,
    13, 20–21
chemical spill, 5–8,
    10, 12, 17
Community Right to
    Know Law, 10, 12
cyanide, 12–14,
    16, 19

I
India, 6, 10–11

M
mine, 12–13, 19

R
Romania, 12–14, 18

T
Tisza River, 14–17

Word Count: 522

Note to Librarians, Teachers, and Parents

If reading is a challenge, Reading Power is a solution! Reading Power is perfect for readers who want high-interest subject matter at an accessible reading level. These fact-filled, photo-illustrated books are designed for readers who want straightforward vocabulary, engaging topics, and a manageable reading experience. With clear picture/text correspondence, leveled Reading Power books put the reader in charge. Now readers have the power to get the information they want and the skills they need in a user-friendly format.